DK Eye Wonder

A PENGUIN COMPANY
LONDON, NEW YORK, MUNICH,
MELBOURNE, AND DELHI

Written and edited by Penelope York
Designed by Cheryl Telfer and Helen Melville

Managing editor Susan Leonard
Managing art editor Cathy Chesson
Jacket design Chris Drew
Picture researcher Marie Osborn
Production Shivani Pandey
DTP designer Almudena Díaz
Consultant Chris Pellant

First published in Great Britain in 2002 by
Dorling Kindersley Limited
80 Strand, London WC2R ORL

2 4 6 8 10 9 7 5 3 1

Copyright © 2002 Dorling Kindersley Limited, London
First paperback edition 2004

Paperback edition ISBN 1-4053-0679-3
Hardback edition ISBN 0-7513-3945-8

Colour reproduction by Colourscan, Singapore
Printed and bound in Italy by L.E.G.O.

see our complete
catalogue at
www.dk.com

Contents

4-5
Where are we?

6-7
Crust to core

8-9
Moving world

10-11
The tips of the world

12-13
The fire mountain

14-15
Earthquake!

16-17
The rock cycle

18-19
Vital survival

20-21
Down to earth

22-23
Nature's sculptures

24-25
Flow of water

26-27
Underworlds

28-29
The power of ice

30-31
The mighty wave

32-33
The ocean floor

34-35
Earth's treasures

36-37
Earth's ingredients

38-39
Ground detectives

40-41
Different worlds

42-43
Planet pollution

44-45
Planet protection

46-47
Glossary

48
Index and
acknowledgements

Where are we?

Where is the Earth? Good question. Let's look into space and find out where we are and what is around us. Then we'll zoom in closer.

Sun Mercury Venus Earth Mars

Let's zoom in on the Earth.

Can you see the towns?

The Earth from space

When we zoom in and take a look at our Earth from space, we can see how the countries and oceans are laid out. You are somewhere down there. This is a photograph of the United States of America taken by a satellite.

Spotting cities

When we look a bit closer we start to see built-up city areas and green country areas. You are now looking at Florida, a state in the USA. Can you see anyone yet?

Jupiter

Saturn

Uranus

Neptune

Pluto

The Solar System

Our Earth is in the middle of a family of planets that all move around our Sun. We call this the Solar System. So far, life has not been discovered on any other planets besides the Earth, but it soon might be!

Hunting down houses

Diving down a bit, we can now see a town in Florida next door to the beach. But we still can't see any people down there.

Finding people

Zoom in on a house and at last, we can see kids! Now look back at the Earth and you'll soon realise how big it is. It's absolutely enormous.

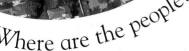

Where are the people?

5

Crust to core

We think we know so much about the Earth and even about space, but what lies beneath our feet? Imagine that the Earth is an apple. The crust that we stand on would be as thick as the apple skin. That leaves a lot of something else underneath.

Journey to the centre of the Earth

Man has only dug about 13 km (8 miles) into the Earth, which is only about a five-hundreth of the journey to the centre. Scientists can only guess what is beneath but we do know that it is very, very hot.

Earth facts

● You may think the Earth is big, but the Sun could swallow up 1,303,600 Earths.

● If you wanted to walk all the way around the Earth along the equator, then it would take you about a whole year, non-stop. You wouldn't even be able to sleep!

All around the Earth is a blanket called the atmosphere, that contains the air we breathe.

The crust is the thin layer of rock that covers the Earth. It can be between 5 and 68 km (3 1/2 and 42 miles) thick.

The mantle is the layer that lies below the crust. The deeper mantle is solid rock but the upper layers are plastic, moving rock.

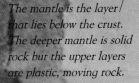

| Granite | Basalt | Peridotite |

The Earth's surface

Earth is made up of rocks. Granite is a typical continental (land) rock. Basalt is a typical ocean floor rock, and peridotite is a mantle rock.

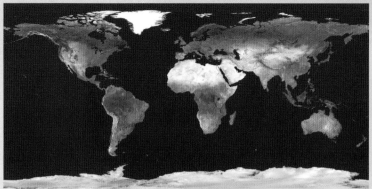

Earth map

About 29% of the Earth's surface is made up of land, which is divided into seven continents (a piece of land that is not broken up by sea). These are North America, South America, Europe, Africa, Australasia, Asia, and Antarctica.

People only live on 12% of the Earth's surface

The outer core is completely liquid. It is made of iron and nickel.

The inner core is a red hot, solid ball of molten iron and nickel that is 4,500°C (8,132°F). That's hot!

Moving world

The Earth's crust is made up of huge plates, which fit together like a jigsaw. The plates have been moving for millions of years and still shift today, with dramatic effects on the shape of our planet's surface.

This is what the continents looked like 200 million years ago.

The continents we know today started to take shape 150 million years ago.

This is the Earth as it is today. What will it look like in another 150 million years?

The continents ride slowly on plates of crust.

Slow progress

The plates drift in certain directions. As they shift, they change in shape and size – this takes many millions of years. See what the Earth looked like 200 million years ago compared

Plate line

The line that two plates run along side by side is called a fault. When the plates move against each other they can create earthquakes, volcanoes, or even mountains.

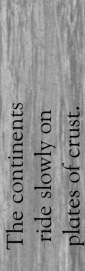

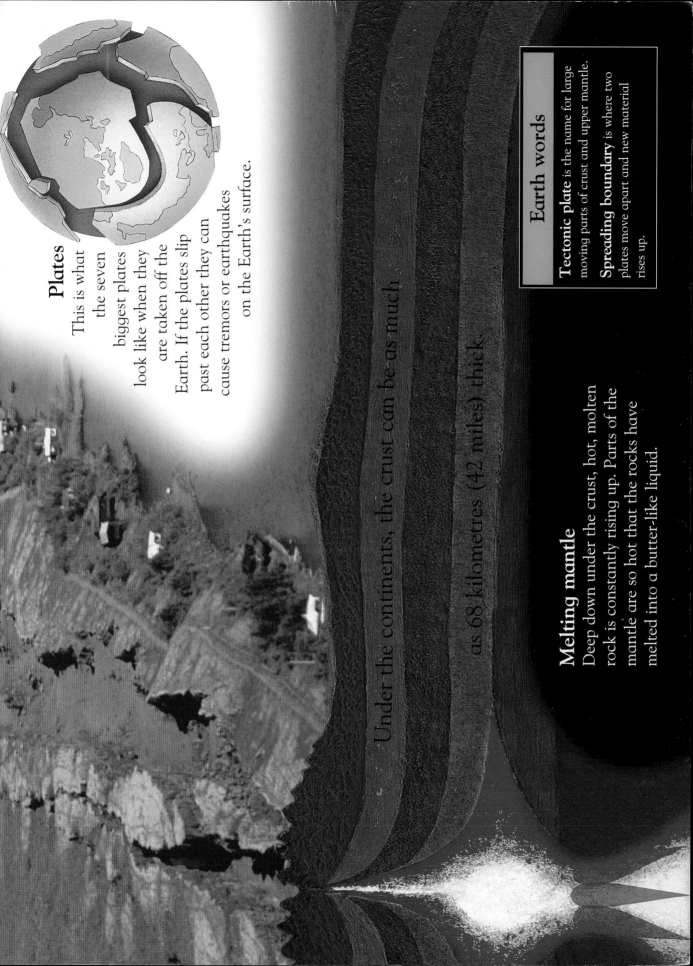

Plates

This is what the seven biggest plates look like when they are taken off the Earth. If the plates slip past each other they can cause tremors or earthquakes on the Earth's surface.

Earth words

Tectonic plate is the name for large moving parts of crust and upper mantle.

Spreading boundary is where two plates move apart and new material rises up.

Under the continents, the crust can be as much

as 68 kilometres (42 miles) thick.

Melting mantle

Deep down under the crust, hot, molten rock is constantly rising up. Parts of the mantle are so hot that the rocks have melted into a butter-like liquid.

The tips of the world

Without mountains, the Earth would look far less spectacular. About 5% of the world's land surface is made up of impressive highland.

Old mountains

Mountains are made when the Earth's crust is pushed up in big folds or forced up or down in blocks. The 450 million year old Scottish Highlands used to be craggy like the Himalayas, below, but wind and rain have worn them down.

New mountains

The Himalayan Mountains, in Asia, are good examples of fold mountains. They are 50 million years old, which is relatively new! Mount Everest in the Himalayas is the highest point on Earth.

The plate pushes forward slowly over the years making more and more folds.

This model shows how plates push together, from the left side, forcing one side to crumple into mountains.

Block mountains

Block mountains are formed when the Earth's crust is moved up or down in blocks. Mount Rundle, Banff National Park, Canada, is a spectacular example of a block mountain.

Fault lines occur and a block drops or lifts to produce a high mountain and a low plain.

Hawaii is the tip of a very, very big mountain.

Underwater mountains

Long lines of islands in the oceans are actually the tips of huge mountain ranges, which lie underwater. The island of Hawaii, Mauna Kea, is the world's tallest mountain from the bottom of the sea to the tip.

The Himalayas are still rising by 4 mm (⅙ in) every year.

The Himalayas began to form when India collided with Asia.

The fire mountain

The pressure builds up underground. Hot, liquid rock, called magma, finds its way to a weak part between the Earth's plates and explodes. Welcome to the volcano.

The big killer

The force of an exploding volcano is enormous – like opening a can of shaken, fizzy drink. Chunks of molten rock as big as houses can be flung high into the air and dust can travel as much as 20 km (13 miles) high.

Mountain makers

As the insides of the Earth explode out of the ground, the lava and ash settle and over time a perfectly shaped mountain is formed. In effect, the Earth is turning a little bit of itself inside out.

The lava that bursts out of a volcano is ten times hotter than boiling water in a kettle.

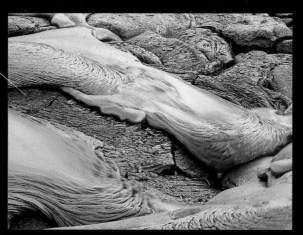

Rivers of fire

When magma pours out of volcanoes it is called lava. It rolls slowly downhill in a huge river, burning everything in its path. When it cools it solidifies into rock, called igneous rock.

KILLER GAS

Sometimes the gas that comes out of a volcano is poisonous. In AD 79, Mount Vesuvius, Italy, erupted violently. A cloud of gas rolled down and poisoned many people in Pompeii, the town at its base. Ash buried them and casts have been made from the spaces the bodies left.

Bubble trouble

In some volcanic areas, you can see heat coming up from under the ground. Mud bubbles and hot water jets, called geysers, shoot up high. They sometimes smell of rotten eggs because of a gas called hydrogen sulphide.

Earthquake!

Imagine waking up one night to find the ground trembling and shaking. That's what it's like to feel an earthquake. These sudden movements in the Earth's plates can cause terrifying damage.

Fault line

The deadly tsunami

When an earthquake happens underwater, vibrations cause ripples in the sea. They grow and grow until they are enormous, deadly waves, or tsunami, that crash onto the shore.

Whose fault?

An earthquake is caused when two of the Earth's plates slide against each other. The line that they slide along is called a fault. When they move they cause vibrations across the ground.

Shock waves caused by an earthquake are recorded by a machine called a seismometer.

Devastation

Earthquakes can be so strong that they cause whole buildings to collapse. Children who live in areas that have earthquakes are drilled regularly on how to remain safe.

The most powerful earthquakes are in Japan.

One in 1923 killed 143,000 people.

The rock cycle

Geologists divide the rocks that make up the Earth's crust into three groups: igneous, sedimentary, and metamorphic. But they all come from the same original material, which moves round in a big cycle.

Igneous rock
Granite and basalt are typical examples of igneous rock. They start their lives as melted rock, such as underground magma and lava that comes out of volcanoes.

You can see the different pieces of sediment in this limestone.

Chalk is also a type

Original rock
Igneous rock either cools down and hardens beneath the surface or on the surface when it erupts from a volcano. It is rock from deep in the Earth's crust.

Break down
Little pieces of igneous rock are broken off by rain and wind and are carried to the sea where they pile up as layers of sediment. The remains of sea creatures are buried in the layers and may become fossils.

Sedimentary rock

Limestone and sandstone are typical sedimentary rocks. You can often see the layers in a big piece of rock when they have been squashed down on the sea bed, such as in this photo of the Painted Desert in Arizona, USA.

of sedimentary rock.

Metamorphic rock

Marble is a good example of this rock. It was once limestone before it was changed by intense heat. When marble is polished it produces beautiful swirly patterns.

This man is chiselling a square piece of marble from a marble quarry in Italy.

Under pressure

When sedimentary or igneous rocks are dragged deep into the Earth's crust by the moving plates, they can get hot, changing them into a new rock type – metamorphic. If the rocks are taken so deep that they melt, they are back where they started as magma. The cycle starts again.

Vital survival

All around the Earth is a protective shield called the atmosphere. It keeps us from burning under the Sun during the day and from freezing at night. Within our atmosphere lie the water and air cycles.

The water cycle

It's incredible to imagine, but the water that we use every day is the same water as was on the Earth millions of years ago. It goes up into the clouds, and back down to Earth as rain, and never stops its cycle.

Water goes up and

Air goes in and out

Water, water everywhere

Water goes up and water comes down. It is evaporated into the atmosphere by the Sun and turns into clouds. When the clouds cool down high up in the sky, rain falls from them.

The air cycle

The air that we breathe is also in a continuous cycle. Animals breathe in a gas called oxygen and breathe out carbon dioxide. All plants take in carbon dioxide and make oxygen.

down, up and down.

and around and around.

Essential air

Because of the air cycle between animals and plants, we could not possibly live without each other. We make the air for each other that is vital for life.

Down to earth

Without soil, life would be impossible as nothing can grow without it. Soil is the part of the Earth that lies between us and the solid bedrock.

Out of the soil grow many plants.

This level is called topsoil. It is rich in food for plants and contains living creatures.

The subsoil has less goodness for plants to feed on.

As you get lower, the soil becomes rockier.

The solid rock below the soil is called bedrock.

Useful soil

Soil can be used in so many ways, from making bricks to providing clay for pottery, but it is most vital for growing plants for us to eat. In South East Asia, hillside terraces stop soil from washing away when it rains.

Layers of soil

If you cut a section through the soil, down to the rock beneath, you would find lots of layers. The material nearest the top is the rich soil needed for plants to grow and the bottom is solid rock.

A handful of soil contains about six billion bacteria!

What is soil?

Soil is made up of rocks, minerals, dead plants and animals, tiny creatures, gases, and water. As plants and animals die, tiny creatures and bacteria break them down to become soil.

Essential food
Plant roots take in water and nutrients from the soil. Plants need these in order to make food and grow.

Wriggling worms
Worms are vital to the soil. They eat decaying plants and animals and deposit them into the soil as they wriggle through it. As they tunnel they help the soil to breathe.

Nature's sculptures

The moment rock is exposed to air, it is in serious danger of being reshaped. Wind, rain, air, and even plants, all seem determined to change the landscape, sometimes with beautiful results.

Frost shattering
This amazing landscape was carved by frost. Water in cracks in the rock freezes and expands and the rock shatters, leaving extraordinary, spiky shapes.

Pillars of the Earth

These strange pillars are called hoodoos. They are formed because soft rock lies below hard rock. Downpours of rain wash away the softer rock, leaving pillars of harder rock above.

Limestone pavement

Limestone is a soft rock that is affected dramatically by rainwater. The slightly acid rainwater changes the limestone into a softer rock, which is washed away. Cracks get larger and the ground becomes uneven.

When air, wind, ice, or plants change the shapes of rock, it is called "weathering".

Watch out! Plant attack

Trees sometimes speed up rock cracking with their roots. As the roots grow, they creep between cracks; when they thicken, they force the cracks to open wider.

Flow of water

Water is incredibly powerful stuff. When there is a lot of it, moving at huge speeds, it can carry away a lot of loose rock and mud. When water changes the shape of a landscape, it is called erosion.

River facts

● The longest river in the world is the Nile in Egypt. Beneath the Nile runs another river deep underground that holds six times as much water.

● The highest waterfall in the world is Angel Falls in Venezuela. The water falls 3 ¾ times the height of the Eiffel Tower in Paris.

Running wild

As water rushes from its source, in the highlands, down to the sea, it constantly picks up chunks of rock, sand, and mud along the way. It then deposits it elsewhere, changing the shape of the land as it goes.

This harder rock is left behind after floods.

Desert floods

Water can even shape the desert. Heavy floods sometimes rush through the land, taking the land with it and leaving weird towers of rock behind, such as in Monument Valley, Arizona, USA.

Water power

The Grand Canyon, USA, is the largest gorge in the world. It has been carved by the Colorado River over 20 million years. Different rocks react in different ways to the water, so the shapes are incredibly spectacular.

Niagara Falls

Niagara Falls is an
enormous waterfall, which
moves backwards by 1 m (3 ft 3 in)
each year. The lower rock is soft
and is worn away by the water.
Eventually the top rock crashes
to the bottom when it can
no longer stay where it is
without support.

Underworlds

Caves can be pretty scary places – dark and damp – but they can also be beautiful. They form when water seeps through cracks in soft rock, such as limestone, and take thousands of years to become caverns.

Most caves are dripping with water.

Gorges form when cave roofs collapse.

Ancient murals

Before people lived in houses, they lived in caves. They painted drawings on the walls like this cattle one. It was painted 17,000 years ago, and found in Lascaux, France.

Cave sport

Caves may be dark, but they are also magical, underground landscapes and some people enjoy exploring them as a hobby. This is known as potholing. It's quite a dangerous sport, however, and must always be done using the right equipment.

The biggest cave in the world, the Sarawak Chamber, Borneo, could fit 800 tennis courts inside it.

It can take thousands of years for stalactites and stalagmites to meet.

Up and down
When water drips from a cave's ceiling, it leaves behind tiny pieces of rock. Over many years it grows downwards into a long icicle shape called a stalactite. Where the water falls to the ground, stalagmites grow up like candles.

The power of ice

There's more to snow and ice than meets the eye. Not only do they produce some of the most spectacular scenes on Earth, but they are powerful tools that sculpt it.

Earth's natural plough

A glacier is an enormous mass of ice that flows downhill slowly. When glaciers melt, they show how much of the Earth has been gorged away. You can see how it has shaped this Norwegian fjord.

The mighty glacier

A glacier is incredibly powerful. It carves its way through mountains, leaving huge gorges or valleys behind. On the way it swallows up and moves giant boulders. Yet it only moves at a speed of about a centimetre or two a day.

Floating island

Some icebergs are huge. But whatever you see above water, there is even more below. Two-thirds lies underwater.

Ice caves

Icebergs are big blocks of ice that break away from the end of a glacier. As they melt, the wind and waves batter them into weird shapes, sometimes creating ice caves.

The mighty wave

When you play on a sandy beach, have you ever noticed how often the waves crash onto it? Well, believe it or not, that wave movement is constantly changing the coastline. Waves are even powerful enough to reshape cliffs!

Waves destroy

Shock waves

● As the waves force coastlines back, sometimes houses built on the cliffs fall into the sea!

● A series of pounding 10 m (33 ft) high storm waves can remove one whole metre (3 ft 3 in) of cliff in one night.

Making a bay

The sea is very persistent. When it finds a weak part along a coast, it breaks through and spreads out as far as it can. It eventually creates a bay, such as Wineglass Bay in Tasmania.

some coasts but make brand new beaches elsewhere.

Creating sand

Sandy beaches take hundreds of years to form. Waves near the shore pummel boulders into pebbles, and with more battering they eventually become the soft, fine-grained sand that you find on a beach.

It's amazing that just water can turn this pebble into fine sand.

Coastline sculpture

This picture shows how powerful waves are. The sea has completely battered its way through the rock on this cliff and formed an archway. Eventually when the arch gets too weak it falls in on itself, leaving stacks behind.

The ocean floor

The ocean is a mysterious place – we can't go beyond certain depths because the pressure will kill us. However, we do know that the ocean floor has some features that are very similar to those found on land.

Coral from surface to 0.5 km (⅓ mile)

Coral reef
Coral reefs are found in clear, warm waters near the shore. Corals are living things and are home to hundreds of others as well.

Earth oceans
More than two-thirds of the Earth is covered in water. The deepest part of the ocean is the Mariana Trench, in the Pacific Ocean, which is 11.5 km (7 miles) deep. Very little life can survive in those depths.

Diving down
It is very difficult for humans or submarines to go down deep underwater. This submarine is called the Nautile and can take three people down to depths of 4 km (2½ miles).

Black smokers
Where the ocean plates move against each other, vents open and hot steam rises into the water. These are called black smokers.

There are many massive abysses on the ocean floor

32

Ocean floor
11.5 km
(7 miles)

Pillow lava
Where volcanoes erupt on
the ocean floor, the lava
solidifies into round, pillow
shapes in the water.

Hatchet fish
Hatchet fish are one of
the few creatures that can
survive in the deepest parts
of the ocean.

Exploding ground
Volcanoes and earthquakes happen
underwater, where the plates meet,
just like they do on the land surface.

Earth's treasures

Hidden deep underground lies a priceless treasure trove of precious minerals, which include rocks, metals, and crystals. We are constantly digging into the Earth to find these minerals as we use them all the time.

Jewel in the crown

Most gems that you find on valuable jewellery start their lives in rock. They begin as crystals but, after cutting and polishing, they end up as beautiful and expensive gems.

Gold diggers

In order to find precious metals and gems, we have to mine for them, and sometimes they are hard to get at. Tonnes of rock, for example, may only hold a few grams of gold.

Gold and silver have been used to make coins and jewellery for thousands of years.

A dash of salt

I bet you wouldn't eat jewellery. Well, gems are crystals and so is the salt that you sprinkle on your food. Pools of sea water are left to evaporate, leaving the salt behind ready to be collected (right).

Mineral facts

● Diamonds are one of the hardest substances. They are used in drill bits, or to cut glass.

● Talcum powder is actually a mineral called talc. It is very soft and crumbles easily.

● Silicon, which is obtained from minerals like quartz, is essential in the making of computers and mobile phones.

Earth's ingredients

Inside the Earth's crust are some essential
ingredients called fossil fuels – coal,
gas, and oil. We use these to provide
energy that runs everything from
cars to the electricity in our homes.

Treasure from the Earth
Believe it or not, coal that we use to burn in
our fires used to be trees that lived 280-345
million years ago. Their remains didn't rot
properly and over time became coal.

Mining for coal
When coal was first discovered, it was quarried
from the surface. Today, there are many very
deep mines, as well as huge opencast
(surface) mines.

Treasure from the sea

Oil and gas were tiny sea creatures
that died millions of years ago and
were buried under the sea floor.
Eventually they became oil. Often,
gas is found just above the oil.

*Sometimes oil is piped straight to the
mainland, but often huge ships called oil
tankers take the oil away from the rigs.*

Oil rigs

A lot of oil and gas
is found in rocks
below the sea bed,
so the only way to
get them is to drill
deep below the sea.
Oil rigs are platforms in
the sea that are specially
built so that we can drill
for oil and gas from them.

Oil facts

● Oil is used for many things,
such as petrol for trains and
planes, and power stations.
Even the ink on this page is
made from oil.

● Workers live on oil rigs for
weeks at a time. Their supplies
are flown in by helicopter.

Sea creatures

Rock

Gas

Oil

*We use fossil
fuels all over our
homes. Some
cookers use gas.*

Rock creatures

Fossils are the remains of plants or animals that have been preserved in rocks over millions of years.

Fossil facts

● Fossils show that starfish lived more than 450 million years ago.

● Dinosaurs appeared about 240 million years ago.

● The oldest fossil is 3,000 million years old and is a microscopic blob-like creature.

Ground detectives

We know a lot about the Earth's history because, amazingly, the Earth tells us all about it. Fossils and layers of rock found deep underground help us to understand the mysteries of the past.

Rock strata

Each layer, or strata, in this cliff in Utah, USA, tells a different story. The top layers are 10 million years old, and the grey areas near the bottom are about 210 million years old. Dinosaur footprints have been found in this layer.

Trapped in time

Amber is fossilized tree resin or gum. Millions of years ago, insects were trapped in amber and remain to this day. Because of this, we have proof that spiders have existed for a long time.

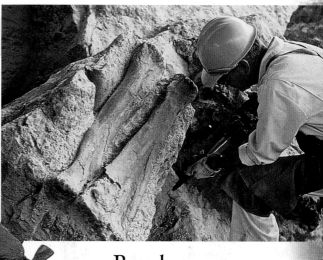

Bare bones

We know that dinosaurs roamed the Earth millions of years ago because we have found many of their bones fossilized into rocks.

By putting all the bones together, it is possible to see what shape the dinosaurs were.

Different worlds

The Earth has so many different climates, weather systems, and types of rock that no one place is the same as another. Animals and plants have had to adapt to each place so they are also very different. Have a look at some of the worlds within our world.

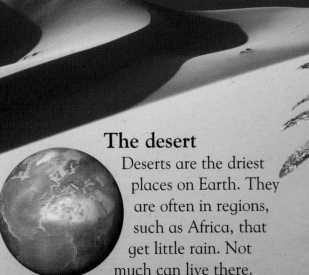

The rainforest

If areas, such as some in South America, have huge amounts of rainfall, then plants can grow. The rainforest has so much rain that the area is covered in plants. There are lots of animals here as there is so much food to eat.

The desert

Deserts are the driest places on Earth. They are often in regions, such as Africa, that get little rain. Not much can live there.

The city

Over thousands of years, people have taken over large areas on Earth and built huge towns and cities on them. A lot of settlements are built near water as it is essential for our survival.

The poles

The top and bottom of the world are the coldest places on Earth, and only a few animals can live there. Penguins thrive in the freezing conditions of Antarctica.

Planet pollution

The Earth is a special but fragile place. Some of the things we do to it, such as polluting it with chemicals, are destroying things that are valuable to its survival. We must learn to look after our home.

Growing deserts

Deserts are expanding all the time. People living near them use up plants for food or fuel, and, once it becomes desert, it is difficult to make anything grow on it again.

Burning forests

Since 1945, more than 40% of the world's tropical rainforest has been cut down for timber and farming. When forests disappear, so do many animal and plant homes.

A load of rubbish

Think about how much rubbish you throw away in your home. Now imagine all the other people's rubbish added to it. It's a big problem.

Polluted air

Factories, especially those that run on fossil fuels, pump out dangerous chemicals into the air that are bad for our lungs. They also make the rain acidic.

A big waste

Each day a huge amount of sewage and chemical waste from factories is pumped into our rivers and seas. Dirty water can spread diseases and can kill fish and other water life.

OIL CRISIS

On March 24, 1989, a huge oil tanker called *Exxon Valdez* had an accident in Alaska. Within a few days, it had spilled almost 50 million litres of oil into the sea. The oil polluted the shoreline and killed a huge amount of fish and birds. It took years to clean up.

Planet protection

How can we protect our planet? Conservation means trying to keep things the same and not destroying habitats by dumping waste. When things get bad, here are some ways that we can help.

The wind makes the turbines spin around really fast, which creates energy.

Rubbish clean up

To cure the rubbish problem, we need to recycle more of our rubbish. Bottles, paper, plastics, and all sorts of other materials can be used again and not buried in the ground.

Wind power

Burning fossil fuels to make electricity puts poisonous gases into the air. Fossil fuels can be replaced with wind turbines. These wind farms are clean and safe for the Earth.

Only about 400 of these beautiful Siberian tigers still live in the wild.

Endangered species

There are many animals, such as this tiger, that have lost their homes and others have been hunted to extinction. Zoos now breed animals and release them back into the wild.

A tree nursery

An excellent way to save the rainforests and to stop deserts from growing is to replant trees as we chop them down. We always need wood, but if we plant new seeds, like these children in India are doing, then we will never run out of trees.

Protection facts

• Car exhausts throw bad chemicals into the air. In the future our cars may run on power from sunlight, called solar power.

• To help clean up rivers and seas, make sure you never drop any litter into them.

Glossary

Here are the meanings of some of the words that are useful to know when learning about the Earth.

Atmosphere the blanket around the Earth that holds in gases.

Bacteria miniature living things, invisible to the eye, that help to convert dead plants and animals back into the soil.

Carbon dioxide an invisible gas in the air that animals breathe out.

Climate the average weather in a particular area.

Coastline the place where land and ocean meet.

Conservation to keep things the same and undamaged.

Continent one of seven huge areas of land on the Earth that are not broken up by sea.

Coral reef a mass of rock-like material that is formed by skeletons on the ocean floor near to the ocean's surface.

Core the hot, central part of the Earth.

Crust the hard outer coating of the Earth that is made from solid rock.

Desert a very dry place that has less than 25 cm (10 inches) of rain a year, which is very little.

Earthquake sudden movements in the Earth's crust that cause the ground to shake violently.

Equator the imaginary circle that passes around the centre of the Earth, between the poles.

Erosion when rock or soil is loosened and transported by glaciers, rivers, wind, and waves.

Fault a break in rocks with movement on each side.

Fossil remains of living things that have been preserved in rocks.

Fossil fuel fuels that include natural gas, oil, or coal, all of which are natural and are formed by dead prehistoric animals or plants.

Glacier mass of ice and snow flowing slowly downhill under its own weight.

Gorge a deep narrow valley cut by a river.

Igneous rock the rock that starts as magma below the surface of the Earth but hardens either underground or on the surface.

Lava red, hot, melted rock that pours out of a volcano when it erupts, and then solidifies.

Limestone a sedimentary rock composed mainly of calcium carbonate.

Magma rock that has melted to a butter-like fluid beneath the Earth's surface.

Mantle the part of the Earth immediately beneath the crust.

Metamorphic rock rock that has been changed by underground heat or weight.

Mineral a simple substance that, either alone or mixed with other minerals, makes up rocks.

Ocean a huge, salty body of water. Also called a sea.

Oxygen an invisible gas in the air that animals breathe in in order to survive.

Planet a large, round object that orbits a star such as our Sun.

Plate a separate section of the Earth's crust that rides on the semi-liquid rock of the mantle.

Pollution materials and gases that are in the wrong place and spoil that environment for the plants and creatures that live there.

Rainforest a tropical forest that receives heavy rainfall and therefore where huge amounts of plants grow.

Rock a large, solid mass underground that is sometimes exposed at the surface of the Earth and is made up of one or more minerals.

Satellite an object in space that revolves around the Earth.

Sedimentary rock rocks formed in layers by the deposition of eroded grains.

Seismometer an instrument that measures the strength of earthquakes.

Sewage rubbish or waste that is carried away in sewers.

Solar power energy that is gained by using the Sun.

Solar system our family of nine planets that revolve around our Sun.

Stack a rock pillar left standing in coastal waters when the top of an arch falls in.

Stalactite a hanging, icicle-shaped structure formed in caves by dripping water with traces of rock in it.

Stalagmite a rising candle-shaped structure formed when stalactites drip to the floor and leave traces of rock behind.

Strata layers of sedimentary rock.

Tsunami a huge, fast-travelling wave that is caused by an underground earthquake.

Volcano where hot magma breaks through the Earth's crust with great pressure.

Weathering the breaking up of rocks by wind, rain, or ice.

Index

Air cycle 18, 19
Arch 31
Atmosphere 6, 18

Basalt 6
Beach 30, 31
Block mountain 11

Cave 26
Coal 36
Conservation 44-45
Coral reef 32
Core 7
Crust 6, 8, 9, 10, 11, 16, 17

Desert 40, 42
Diamond 35
Dinosaur 38, 39

Earthquake 8, 9, 14-15, 33
Electricity 42, 44
Endangered species 44
Equator 6
Erosion 24

Fault 8, 11, 14
Fjord 28
Fold mountain 10
Fossil 16, 38

Fossil fuel 36-37, 43, 44
Frost shattering 22

Gas 36, 37
Gem 34
Geyser 13
Glacier 10, 28, 29
Gold 34
Gorge 24, 26, 28
Granite 6, 16

Hoodoo 23

Ice 28-29
Iceberg 29

Igneous rock 13, 16, 17

Lava 12, 13, 33

Magma 12, 13, 16, 17
Mantle, 6, 9
Map 7
Metamorphic rock 16, 17
Mineral 20, 34
Mountain 8, 10-11
Mural 26

Ocean 11, 32-33
Oil 36, 37, 43

Peridotite 6
Planet 5
Plate 8, 9, 10, 12, 14, 17
Pollution 42-43

Rainforest 40, 42, 45
River 24, 28, 45

Salt 34
Sand 30
Satellite 4
Sedimentary rock 16, 17
Seismometer 14
Soil 20
Solar System 5
Space 4, 6
Stack 31
Stalactite 27
Stalagmite 27

Tsunami 14

Volcano 8, 12-13, 16, 33

Water cycle 18
Waterfall 24, 25
Wave 14, 30-31
Weathering 23

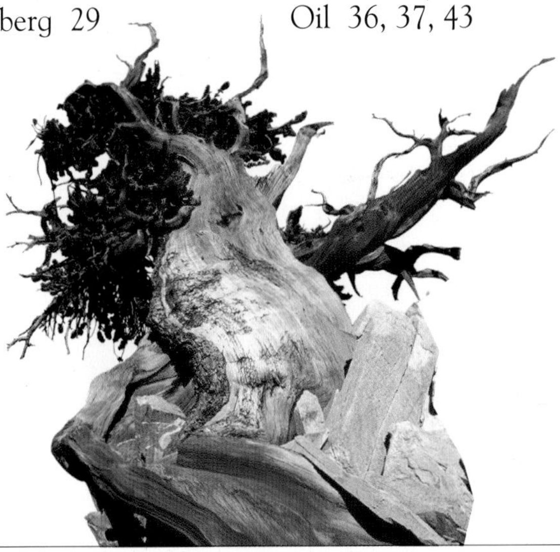

Acknowledgements

Dorling Kindersley would like to thank:
Dorian Spencer Davies for original illustrations;
Jonathan Brooks for picture library services.

Picture credits:
The publisher would like to thank the following for their kind permission to reproduce their photographs:
a=above; c=centre; b=below; l=left; r=right; t=top;

Ardea London Ltd: Graham Robertson 39l. **Bruce Coleman Ltd:** 19t, 40tr, 40br, 44bl; Jules Cowan 22; Pacific Stock 32-3. **Corbis:** 14-5, 21, 23tl, 28-9b, 28-9t, 28r, 29tl, 29r, 32c, 46, 47, 48 boarder; Archivo Iconograofico 26c; Hubert Sadler 16l; Stuart Westmorland 4-5b; Ted Spiegel 17r; Tom Bean 16-7. **Environmental Images:** Herbert Giradet 42tl. Gables:27r. GeoScience Features **Picture Library:** 39cr. **Robert Harding Picture Library:** 10tr, 16cl; Thomas Laird 10c. **Hutchison Library:** 40main. **The Image Bank/Getty Images:** 1, 4cr, 24c, 37r. **Masterfile UK:** 23br, 36c, 40l, 48c; John Foster 24l.

N.A.S.A.: 4-5t, 7t. **Natural History Museum:** 17tr. **N.H.P.A.:** Anthony Bannister 34tl; Haroldo Palo Jr. & Alberto Nardi 26-7; Robert Thomas 26bl. **Planetary Visions:** 2, 18-9, 40tc, 40cr, 40bl, 41tc. **Powerstock Photolibrary:** 3, 18b. Rafn Hafnfjord: 8-9. **Science Photo Library:** 4-5ca, 12-3, 20l; B. Murton 32-3b, 33tl; Bernhard Edmaier 12l; David Nunuk 23tr; ESA 4cl; G. Brad Lewis 13t; Martin Bond 25, 30. **Still Pictures:** 34c, 34-5, 36bl, 41r, 42cr, 42b, 43, 44cl, 45. **Corbis Stock Market:** 24bc. **Stone/Getty Images:** 11tl, 14l, 19b, 30-1, 31t. **Telegraph Colour Library/Getty Images:** 5br, 14c, 20c.

Jacket Images: Natural History Museum: spine b. **Still Pictures:** front b. **Stone/Getty Images:** back, front top.

All other images: © Dorling Kindersley. For further information see **www.dkimages.com**